HonorSociety.org Reddit Review

The Beginner's Guide to Success on Reddit

Mike Moradian

Copyright © 2020 by Mike Moradian

All rights reserved. No part of this book may be used or reproduced by any means, graphic, electronic, or mechanical, including photocopying, recording, taping, or by any information storage retrieval system, without the written permission of the publisher except in the case of brief quotations embodied in critical articles and reviews.

CONTENTS

Chapter 1: The History of Reddit 3
Chapter 2: Understanding Subreddits 3
Chapter 3: Voting 3
Chapter 4: Posting Content 3
Chapter 5: Lingo & Formatting Norms 3
Chapter 6: Who is Not Welcome on Reddit? 3
Chapter 7: Reddiquette 3
Chapter 8: 5 Reasons to Join Reddit Today 3
Chapter 9: The Future of Reddit 3
Conclusion 3
References 3
Introduction 1

Introduction

We all know the big name social media platforms today. We know of Facebook, Instagram, Twitter, Snapchat, and now of course, TikTok. Many of you reading this have, at the very least, a Facebook and an Instagram. It's part of the norm in 2020.

What many people don't know is that there are dozens of "social media sites" at large, churning along with millions of users right under your nose. There is perhaps no internet trove that encapsulates this kind of hidden participation better than Reddit.

Reddit is the dark horse in the world of social media. Many of you reading this might not have heard of it; or, if you have, you are probably unsure of how it works. That's how Redditors like it – they like having this little slice of the internet that is not easy to follow, packed with specialized lingo, and wholly unwelcoming of marketers, advertisers, and self-promotion (it's like the antithesis of Instagram).

Created in 2005, during the arrival of Facebook and most other major social media platforms, Reddit has been playing the game all along – you just might not have noticed. Today, the discussion platform boasts over 330 million worldwide users, with 26.4 million Americans that use the discussion site monthly. It has undoubtedly claimed the title as the number one news, information, and discussion social media site where users feel free to share information, leave reviews, explore niche topics, and go down the dark internet rabbit hole regarding some pretty insane ideas.

Reddit isn't necessarily made for everyone. It's beloved by those that want to discuss theories, posit new ideas, challenge norms, and strike up conversation not necessarily appropriate at dinner. However, you can find everything from how-to guides regarding new crafts at home, to evidence of alien invasions throughout human history.

Therefore, it's important to understand this platform with greater clarity today. It is both a place to meet people and engage in debate, as well as further your understanding and knowledge of a topic (many times without the bias we see in aggregated social media timelines today).

In this Reddit guide, we are going to look at the basics of the app, the lingo, voting procedures, content posting, formatting, and Reddiquette, as well as those not welcome on Reddit and how you can master the platform by the time you're done with this book.

So without further ado, it's time to dive into *The Beginner's Guide to Success on Reddit*.

CHAPTER 1:

The History of Reddit

In order to fully understand the Reddit we know and love today, it's important to first understand where this platform has come from, what it has endured, and what it has changed to arrive at where it sits today.

As we have mentioned, this platform was first launched in 2005 – about 15-years ago. Steve Huffan and Alexis Ohanian, two 22-year-old University of Virginia graduates, together founded Reddit after receiving funding ($100,000) from Y Combinator in June 2005. At the time, the two founders were submitting most of the links from many fake accounts to make the site look like it was popular and engaging.

In 2006, they sold Reddit to Conde Nast Publications for $20 million USD, as the site was receiving 500,000 unique daily views at the time. Proceeding to operate as a subsidiary of Conde Nast's parent company, Advance Publications, by 2011, Reddit outranked Digg, a major competitor that had dominated the online discussion space.

In 2009, Steve Huffman left Reddit, citing vision differences and the desire to pursue something else. In 2010, Alexis left Reddit and joined Steve to found Hipmunk.

In 2011, following Reddit's major ranking success, the company spun out as its own entity again. When the company spun out, Yishan Wong was appointed as CEO. Reddit was starting to pick up some serious steam at the time, which also meant it was at the center of some stressful controversies, from a nude celebrity photo leak and hate speech/gun sale problems, to funding issues that forced the team to relocate to San Francisco following a big funding win. Wong was, of course, criticized for how he handled most of these issues, leading him to appear "noticeably" stressed to everyone working with Reddit at the time.

Resigning the following year amid calls for him to step down, Wong said his intention was to protect the news site he had loved since its early days. He took to Quora the next day and said, "I am basically, completely worn out."

Ellen Pao was announced to take over at the time, with Alexis returning as an executive chairman to assist the newly independent company again. It was chaotic, to say the least, with former CEO Huffman stating at the time, "it wouldn't be Reddit if it was functional."

Many news outlets at the time ran stories regarding the success of Reddit, yet commenting on its inability to convert its popularity into a viable business model. "Traffic was never the problem, everything else was," wrote Mashable.

Everything seemed amicable for some years until Pao resigned as CEO in 2015. Steve Huffman was there in the waiting, ready to snatch up his leadership position again. Huffman is currently the CEO of Reddit, working closely with Alexis to execute their vision yet again for the platform.

Now that you have a better understanding of the tumultuous past that gave rise to the Reddit we access today, it's time to dive into some basic information for better understanding this internet movement.

What is Reddit?

One of Reddit's biggest branding problems is that it can be defined as so many different things to people. There's no "one thing" that Reddit does. In its simplest form, Reddit is a social news site that enables users, or Redditors, to create accounts, post content, and comment on other people's content. When someone goes to share content, they have two options: they can post a link to an external website, or they can write a text "self" post. Each post is open to the Reddit community and can receive up and down votes (we will look at this in coming chapters).

As opposed to Instagram or Snapchat, which is an entirely self-centric social media experience, Reddit is about community. If the entire community feels certain content is good or bad, they can take action on that content. Therefore, when posting to Reddit, Redditors are supposed to think about others and their impact on these people – as opposed to just posting a photo that shows off one's curves, etc.

Reddit has maintained a 90s looking, retro-style website that the users have come to adore today. It's not minimal, white, or filled with nice colors and rounded circles. Rather, the Reddit color-scheme is neutral, monotone, dark, and reminiscent of those Dell computers we all had in our basements during the 90s. On Reddit, you can find trends and memes, plus the space alien mascot that has come to define the site in recent times. You can also find celebrity Q&As, as well as other threads that make it easier for users to come into direct contact with the internet heroes.

Which leads to the Reddit slogan that everyone loves to say today: Reddit is the front page of the Internet. Reddit takes the features of a forum, social network, and news aggregator, and plugs it all into one. It's something like a crossover between Twitter, Quora, Tumblr, and every other chatroom on the internet where users share information about common interests. That's why it's one of the most popular websites where people can upload stories, links, and content, sharing it with a greater community of Reddit lovers and adorers.

Reddit is Going to Seem Foreign

At first, you won't speak the language, you won't understand the customs, and you'll accidentally upset or offend some Redditors when you get onto the site. It's important to know that right now so it doesn't stress you out or deter you from trying it out fully. Like anything new, it's best if you get on the site every day and watch how content is moved around, posted, and commented on. Don't just look at outbound links – dig into comment threads and really get a feel for what people like to chat about.

Here are some other newbie Reddit tips:

- **Redditors Refer to the Site A Lot:** If you see something posted that just plain doesn't make sense to you, it's likely referencing an aspect of Reddit culture, which can be a meme or something else that is unique to the site. Generally, if you keep reading in the comments, you'll be able to find a reference point that helps you better understand what the comment means. When someone does a good job of explaining things, other Redditors will up vote their explanation so its easier for you to find at the top of the thread.
- **Comments Tell All:** Reddit is a platform where you are free to post anything you want. This has gotten them into trouble at times, but it remains central to the site. Therefore, if someone starts a subreddit with incorrect information, the comments will tell you that immediately. If the thread has been there long enough, there's a good chance the best contradictory comment is sitting at the top of the thread. This is how the community self-regulates content without the intervention of Reddit personnel. Facebook and Twitter need external intervention, etc. to clean up things.
- **It's Going to Take Time:** Like TikTok, you won't be able to figure out everything in one

day. It's important to approach Reddit with patience so you are not frustrated or likely to quit it. Take one week to observe, read, and dig deep. You will finally understand it, and probably be hooked.

Reddit Facts

Some important things to note today: the Reddit user base is primarily male. The average user spends about <u>16 minutes and 10 seconds</u> on Reddit per day, with Canadians being the biggest "time wasters" on the site today.

Some more stats to takeaway:

- 150 million pages are viewed on Reddit every day
- There are 1.7 billion comments on Reddit
- Reddit is the 6th most visited website in the United States
- Reddit is the 7th most visited website in the world
- There are 40 million searches made on Reddit every day
- More people use Reddit than Snapchat every day
- More people use Reddit than Twitter every day

At this point, you should have received the following impressions: Reddit was created in chaos and in some capacity, still functions that way today; Reddit is the antithesis of common social media sites today and is actually more accessed than some of the biggest apps of all time, like Twitter; and that Reddit is perfectly happy being its own, unique, dysfunctional family that enables community members to post, comment, and vote on content every day.

With that information, it's time to explore the nuances of this site and how you can "blend in" as a Reddit user if you are just getting started.

CHAPTER 2:

Understanding Subreddits

T he first Reddit lingo term we're going to introduce to you in this e-book is what's known as a subreddit. It's these kind of unique terms and concepts that can make Reddit so overwhelming if you sign on with no background information.

Therefore, let's start from the beginning.

To use Reddit, you will need to sign up for an account through a quick and self-explanatory process. You will make a username, profile, password, and so forth. When you are done with this process, Reddit will prompt you to add some subreddits to your account.

What is a Subreddit?

As subreddit is an individual message board devoted to one particular topic. They are represented by the letter "r" with a dash following it. For example, there is a thread called r/news and r/showerthoughts. Each subreddit is hosted by a moderator. That moderator gets to establish its own rules for what content can be shared, what content can't be shared, what can be commented, and so forth. If you do not like the rules associated with a subreddit, you can use the search feature to find another subreddit that is more appealing to your communication style.

To find subreddits that interest you, simply enter a topic into the search box that is located in the upper right-hand corner of the homepage. If you can think it up, chances are, there's a subreddit for it. From Game of Thrones fan theories, to information the government is not telling us, someone, somewhere, has made a subreddit for it.

Once you search this topic, you'll be taken to a page with a list of suggested subreddits associated with that topic. If you see one that catches your eye, you can click on the green "subscribe" button. This ensures you will be notified when contributions or changes to the subreddit occur.

When you want to return to the home page from a subreddit, you click on the little alien figure in the top left. However, your home page will now display posts from that subreddit. It's similar to the layout of Tweetdeck – a feature that enables social media marketers to view Twitter feeds from multiple accounts.

As you add more subreddits, your home page will expand and grow in depth, displaying all of the subreddits to which you are subscribed. Additionally, as you subscribe to these subreddits, you will see them listed at the top of your Reddit homepage for easy organization. If you want to dive into one particular subreddit, you can click on its name at the top of that page to open it up. When you go back into the subreddit, posts will be arranged according to how many up votes they have received (we will look at this in the next chapter). However, Reddit will rank more recently contributed posts at the top in order to give them a chance to be seen and therefore voted on.

By providing new posts with exposure, it gives Redditors a chance to have their posts blow up. Like any social media site, users want to know their content has the ability to go far and wide, influencing people and changing opinions. It's every Redditor's dream come true to have one of their subreddit posts "blow up" and receive serious voting traction.

The Front Page

Reddit has its own home page for all worldwide users separate from your subreddits. The home page consists of content that has been submitted to the most popular subreddits on the site, like pictures, technology, world news, music, gaming, and so forth. New users are automatically subscribed to these subreddits to help them get a feel for the content available to them today. By perusing these massive threads, you can learn more about subreddits that are niched and catered to your liking.

When subreddits gain enough subscribers, it can become part of the default home page for all of Reddit, exponentially increasing its traffic. Subreddits that recently joined the home page include r/atheism and r/minecraft.

If there is content on your home page that you no longer wish to see, you can simply unsubscribe from the subreddit to clean up the appearance. Click on the red button at the top right corner of that community.

What Are Multireddits?

Multireddits are customizable groupings of subreddits per your choosing. This enables you to navigate the communities of your choosing without having to look at the front page every time. These groupings will appear on the expandable sidebar within your Reddit. They can be created by selecting the "create" button and then entering the subreddits of your choosing. You can even get a peek at other public multireddits, and if you like what you see, create a copy of their curation for your own sidebar.

By mastering the art of multireddits, you can more efficiently gather the information you seek every day when you sign onto the platform.

CHAPTER 3:

Voting

Voting is where Reddit starts to really show its true colors. The community-mentality of Reddit comes into the picture when members can decide if something deserves an up or down vote. Voting is the idea that the community can choose if something sinks or swims. If you like something you are reading, or you like a post someone put in a subreddit, you can up vote that post to show your support. However, if a post doesn't really make you happy, or you feel its inaccurate, wrong, etc., you are free to down vote it as well.

How does this work?

To the left of every post, you will see a grey number with an arrow that points above or below it. The number you see represents the post's rank – which is configured by the amount of up or down votes. You can apply votes to comments within subreddits as well. The same arrows will be available next to every comment beneath the post, too.

Fuzzing Scores

Reddit has admitted to altering certain post's scores from time to time in order to prevent spam and abuse. If someone is trending for a post about child pornography, or if a spam account is boosting its own score, Reddit will intervene to try and make the community safer for everyone. This started following the fiascos in 2012 when dangerous gun subreddits, etc. were promoting violence.

You will also notice that at times, posts with the highest score will not be at the top of the subreddit. This is to allow new posts to have their fair share of time to rank well. Reddit would not be as fun if the same posts ranked at the top of every thread, every single day. This also gives users the incentive to keep commenting on other threads so they can have their new comments go viral.

Reddit refers to this phenomenon as their time decay algorithm. The decay holds that a 12-hour-old post must have 10-times as many points as a brand new post to appear in similar ranks. Any given story also has a lifespan of max, 24-hours, on a user's home page. Much like a news site, Reddit wants to churn out brand new content every day so that users have incentive to log into their accounts every day as well.

Sorting Submissions

Seeing what's "hot" is always fun, but so is seeing what's "fresh" and in the backend of Reddit as well. If you want to see brand new submissions that haven't even had time to gain traction yet, click on the "new" tab at the top of the front page, or choose a subreddit. There you can see things that happened just moments ago. You can then up vote or down vote them, as well as comment on them, playing a role in where that subreddit is going to go.

Most of the content in this section can be uninspiring, but every so often, that golden thought comes through that will make you happy you were the first one there to read it.

Rising Links

You can also see which posts are picking up traction by viewing the "rising" link category at the top of your home page. This section will show you posts within a new tab that are gaining a lot of traction, but haven't quite hit that home page yet.

Controversial Tab

Of course, everyone on Reddit is there to get involved with a little controversy. That's what makes the site so fun – nothing is off limits. If you want to see which posts are batting for a score that is more positive or negative, you can click the controversial tab at the top of your home page. Play a part in determining which way the thread is going to go.

What is Karma?

Although this can also be included in the Reddit lingo portion of this book, it's appropriate to mention in the voting chapter. Reddit Karma is the accumulation of "goodwill" when your post or comment receives enough up votes to be ranked as Karma. This doesn't give you any influence or anyone else, nor can it be used to get money or any special access. It does, however, tell other Redditors when they visit your profile that you add some serious value to the community.

There are two kinds of karma up for grabs on Reddit: one for links and one for comments. Both types of karma are displayed on your user profile in the upper right-hand corner. They are to be seen as "badges of honor" or the highest form of Reddit accreditation available today.

Now, it's time to look at the process for sharing your very own, personalized content and commentary today.

CHAPTER 4:

Posting Content

After you spend time on Reddit reading posts, observing comments, subscribing to subreddits, and so forth, you will reach a point in which you want to post your own content. That's the name of the game, right? Be aware that your first time posting content can feel intimidating – what if it gets voted down immediately? It's a fair thing to wonder and remember – no one is an expert at something they are new to.

Be confident in your thoughts and commit yourself to sharing them.

The Posting Process

Posting content to Reddit is actually very easy, since the platform is committed to that retro 90s appearance. When you go to submit to Reddit, you must first choose if you want to write a text post or include a link.

Next, you need to pick a title. Try and keep the title between 4-9 words. However, some titles may need to be longer or shorter on purpose. If you are posting a link, you will then paste the URL in the open box below. There will be a button called "suggest title," which will actually aggregate title ideas for you based on your link.

Next, you will choose a subreddit to post the comment or link to. Reddit will list "popular choices" below if you are having problem coming up with one. If not, you can type the subreddit title right into the bar. Click submit, sit back, and see if your comment rockets to the front page.

Remember: more than 70,000 links are shared to Reddit per day, which means the likelihood of yours picking up steam is going to be slim. But, like any social media site, the challenge can also make it that much more fun and engaging.

Tips for Viral Comments/Posts

Here are some helpful tips that will help you create content people want to share and read:

1. **Cater Your Content to Your Community:** If you are posting in a gaming subreddit and go off on a tangent about your bunny rabbit with pictures of your pets, people are going to boot you out immediately. You need to think about the community you are targeting. If you have a pet bunny, posting to the r/rabbits (2,767 subscriber) or the r/aww (864,407 subscriber)

subreddit will make it much easier for you to get traction with the content. A subreddit as small as r/rabbits will make it almost effortless to go to the top of the thread after you post. But, since the thread is small, the post will never make it to the home page.

2. **Bring Out the Wit:** People are on Reddit to be informed, but also to be entertained. If it wasn't entertaining, they'd find something else to do with their time. Therefore, the majority of posts on Reddit are funny, witty, and relevant. Funny pictures, stories, memes, web culture references, and self-referential humor are the norm. If you are new to the community, a lot of this banter might not make sense yet, which is why you'll want to learn the big lingo terms we're providing in the next chapter.

3. **Say Hi to Politics:** If you want to get down and dirty with political discussions, Reddit is the place for you. You can participate in conversations, post political news headlines, and get into some gnarly fights with other Reddit members. The site is skewed to the liberal side, bordering on libertarian for many. However, there's no shortage of conservative and moderate voices on there trying to balance out the conversation. Controversial political

topics often make their way right onto the front page.

4. **People Want More Knowledge:** One of the main reasons why people end up on Reddit is that they want to know more about a particular topic. It's regarded as an amazing place to share news and engaging in intellectual rhetoric. Therefore, interesting facts, first-hand personal stories, and so forth are very welcomed on Reddit. There is a "Today I Learned (TIL)" subreddit that is used for sharing eye-opening facts, as well as fascinating factoids and so forth. If you're an information junkie, Reddit is the place for you.

Reviewing Content Similar to Your Own

If you want to share an article or a concept from a website and you want to know how many times it has already been shared on Reddit, you can search the domain link in the Reddit search bar. Paste it in and click enter. You will be shown a number of results for content from that specific domain, allowing you to review the various subreddits to determine where the content has been regularly sent to.

Once you find a few subreddits that are keen on that domain and that kind of research, you can update your search to add in the subreddit specific parameter. This looks like: subreddit:nameofsubreddit. This enables you to see the selected content that was just sent to that specific subreddit so you can be in the know before you submit it as well.

How this looks in practice. You look up searchenginejournal.com. You see a lot of that URL is mentioned in the Technology Subreddit. You can then search the URL plus subreddit:technology to see that domain specifically in that subreddit.

How to Submit the Very Best Content

Now that you know what you need to know as well as which subreddits to follow, you need to consider what performs best in that subreddit when it's your turn to contribute. Click on the TOP tab in the subreddit to view all of the comments that have ranked at the top. You have the choice of filtering for the time period as well. Use the domain search trick above to see how your potential links are going to perform based on how they did in the past.

Now it's time to learn the ins and outs of Reddit lingo.

CHAPTER 5:

Lingo & Formatting Norms

Y ou need to understand there is a whole world of Reddit lingo that is not going to make sense to you are first – and that's ok. There are persistent acronyms that might leave you confused, which is why we want to clear the major ones up here so you can be prepared.

- **Original Poster (OP):** If you see the acronym OP used, it is referring to the original poster or subreddit creator. "According to the OP, this image was removed from Times Square yesterday." It's an easy way to accredit the original moderator in the discussion moving forward.
- **TIL:** No, TIL does not mean "until" in slang on Reddit. TIL means "today I learned." For example, "TIL that the Easter Island statues must have been created by an outside life force. There's no way the Polynesians had the machinery to make those monstrosities."
- **DAE:** DAE means "does anyone else…" You would use it in the following, "DAE think that

Ellen DeGeneres is the funniest TV talk show host on television right now?"

- **IAmA:** This one might not make much sense to you because it doesn't actually shorten a phrase. It just means "I am a..." It's a cutesy way of writing it, using it say "IAmA student in New York City." It's one of those Gen Z kind of communication norms that is becoming more common today.

- **AMA:** You might have seen this acronym on other sites as it is not just unique to Reddit. AMA stands for "ask me anything." You could start a subreddit on your expertise in epidemiology and post, "I just received my PhD in infectious diseases today. AMA."

- **TL;DR:** This is when Reddit can get sassy. Someone will use this expression to say, "too long; didn't read." Sometimes on Reddit, people can write entire essays in their response. This caters to users that want to soak up every last piece of information. However, if you don't want to spend 20-minutes reading a response, you can reply, "TL;DR I think that Vegas is the best city in the U.S. because I found a monkey in my hotel room when I went there. Just saying."

- **FTFY:** This is a cute acronym that pays homage to the community mindset on Reddit. It stands for, "fixed that for you." You can use it if you change an image or make a meme out of another person's post. It's typically used in a comical setting.

- **Brave:** Most times, if someone calls you "brave" or writes "so brave," they are being sarcastic. This is their way of saying whatever you have posted is "so brave of you."
- **Cakeday:** Cakedays aren't Reddit users' real birthdays; rather, they are their Reddit birthdays or the day they joined the site. If you see a little cake icon next to someone's username, it means they are celebrating their cakeday. It's just another way to further the feeling of community on the site.
- **Crosspost:** A crosspost refers to a post that has been posted on more than one subreddit.
- **Edit:** It's common Reddit courtesy to post edit if you have changed your original post or comment, as well as explain why.
- **ELI5:** This acronym stands for "explain it like I'm five years old." This is a simple way to ask someone to dumb down what they are trying to say.
- **FTA:** FTA refers to "from that article," which is an easy way to reference another piece of writing in a post.
- **Hivemind:** This simply refers to the overall community of Reddit and its beliefs.
- **ITT:** ITT means "in this thread."
- **Karmawhore:** This is a derogatory term used to describe Redditors that comment and add links designed to pander to Reddit users for more karma.

- **NSFW:** This acronym is used all over the internet today. It refers to "not safe for work," meaning, don't open the picture or link if you are sitting in an office.

Using a Pseudonym

Did you know that just about every Reddit user, including the founders, goes by a pseudonym on the site? Unless you are using Reddit for a branded account, running ads, or managing a profile, you should use a pseudonym when creating your username. And, even if you do make a branded account, it's recommended to make a second pseudonym account that will engage with your branded account to get the activity increasing.

Yes, Reddit allows you to make multiple accounts.

CHAPTER 6:

Who is Not Welcome on Reddit?

As [Mashable so aptly stated](): there is a zero tolerance policy on Reddit for marketers, fakers, and spammers.

Posts designed to self promote, spread fake news, or spam the site will be reported and removed very quickly. Any activity is down voted out of site before anyone working at Reddit even has a chance to find it. Redditors take great pride in their authentic community, and they have no problem acting quickly to protect their authenticity.

Self-Spammers

Submitting your own content is pretty obvious on Reddit. Reddit follows a 9:1 rule when it comes to this – about 10% of your submissions are allowed to be your own content for self promotion. However, keep it light and don't do it frequently.

This, of course, refers to commenting on other threads. You are free to supply your own content on your page and threads as much as you want.

Non-Commenters

Commenting is a massive part of being a Redditor. In fact, if you do not comment ever through your account, the platform will flag you as a spam risk. If you want to fit into the community, after your first week of observing, it is highly recommended that you comment and start voting on comments. That is how Reddit stays alive and thriving, every single day.

This holds true for failing to respond to comments on your thread. Reddit is meant to be interactive – it's not like Facebook where fights break out under posts. Debate is part of Reddit and welcomed. When someone writes under your thread backing you up or arguing with you, you are expected to respond to them. That's the whole point of Reddit.

Off Topic Posters

As we've mentioned before, if you throw down some comment about a totally random topic related to a subreddit thread, they are not going to be happy with you. Try and stick with the niche, and if you have something to say, try and find a subreddit that matches that topic.

Over-Enthusiastic Posters

Yes, it's possible to submit too much to a subreddit. Regardless of how high quality your responses and links might be, the creator of that subreddit wants other people to have an opportunity to contribute their thoughts as well.

If you are posting about topics or theories that are continuously down voted out of a subreddit and keep pushing that narrative under the thread, people are going to get annoyed.

Every Redditor is supposed to respect the greater consensus. If the majority of people are down voting something you are saying, you are supposed to back away.

Copy and Pasting to Multiple Subreddits

Subreddits are supposed to be their own unique incubators. Therefore, it is not appreciated when people type up a concept, perhaps related to macroeconomics, and paste it into 10 different subreddits that revolve around economics and trade. Many of the subreddit subscribers are subscribed to the other groups as well. They won't like this and they'll flag you and your account as spam to Reddit.

CHAPTER 7:

Reddiquette

Reddiquette is a cute and clever name that refers to the basic rules and decorum of the Reddit community. If you are unsure of how to behave on Reddit or what you should be doing to please other members, Reddit so kindly contains its own FAQ that can answer these questions for you.

One of the most important elements of Reddiquette is to avoid self-promotion on the platform. At this point, you absolutely know this to be true and therefore are figuring that you will need to be cleverer in your marketing tactics – rightfully so.

The Original Source

Another important element of a true Redditor is finding the source of information and using that link as opposed to a blog that just talks about the research. This minimizes the spreading of fake news, as well as prevents people from posting self promotional blogs and articles about themselves.

Resist the Spamming Urge

There are a lot of guides out there that claim they can help you hack Reddit to end up on the first page. However, Reddit has taken a strong hand when it comes to fighting spam on their site today, which is why trying to game the system is not worth that 24-hour moment of fame on the front page. They will ban your account forever, and they'll ban anyone else associated with what you did to get on the first page.

What is Reddit Gold?

Naturally, as a community-oriented online experience, there is of course some kind of community currency exchanged on Reddit. Known as Reddit gold, you can purchase this valueless currency to unlock new features on Reddit, or support the company as a whole. If you want, you can buy gold for yourself, gift it to other users for particularly good posts, or sit on it until you're ready to use it for future posts.

Here are some things you can do with your Reddit gold:

- **You can elect to turn off ads during your time on Reddit by applying your Reddit gold.** This will ensure those banner ads across the top of the site are no longer there.

- **If the general Reddit template is starting to bore you, Reddit gold will help you buy themes and designs.** You can choose between different themes to personalize your reading experience.

- **Reddit gold enables you to highlight where new comments are since the last time you posted.** If you don't want to lose your place next time, a few gold coins can do the trick.

- **Available only on desktop, Reddit gold helps users keep track of the links they've visited through Reddit, no matter what computer they are using.**

Secret Marketing Wins

For those of you interested in somehow using Reddit to market, here is one example of a company that figured out how to make it happen. Degree, the antiperspirant company we all know and love, dove deep into a meme based on survival show star Bear Grylls. They took to Reddit and hosted an AMA with Bear where he answered questions about wilderness survival through YouTube videos hosted on a site called the Adrenalist.

Louis C.K. and Aziz Ansari have also done AMAs exclusive to the Reddit community. In these instances, the Reddit community welcomed the AMA because they provide new intel and information to the community without blatantly promoting any product.

However, it doesn't always end well. When Woody Harrelson did an AMA to promote his forthcoming movie, he ignored some of the harder questions that were submitted, which caused the Reddit community to turn on him.

Therefore, if you're going to market with an AMA, market yourself and your experiences, not a movie or a particular product.

Additionally, always be transparent. The second Redditors think they smell a rat or a liar, they will vote you out of existence.

You Earn Your Place on Reddit

If you are active, engaged, honest, open, and transparent on Reddit, you will have no problems. It's all about pursuing the truth, and nothing but the truth. Anyone that appears to have an agenda or stand in the way of pure, unreasoned fact will be taken down. If you are telling the truth and genuinely excited about conversations, the world of Reddit will be happy to welcome you in.

But: never forget to read the subreddit rules. They can vary – and they are important.

CHAPTER 8:

5 Reasons to Join Reddit Today

N ow that you are aware of the good, the bad, and the ugly when it comes to Reddit, we are going to posit why you should still join this amazing social media market today. As the dark horse in a race of social power, there are many reasons why Reddit can totally change your personal and professional life.

Here are just a few:

1. Research and Decision Making:

Let's say you need to make an important decision. Maybe it's to choose between colleges, or maybe it's to find the real source of an article. You don't know what to pick and you just can't find what you need on Google. That's where Reddit comes into the picture. Reddit is teeming with subreddits that have the information you need and desire. There's a subreddit for everything. Maybe you want to buy a new book and you're down to two options. Someone, somewhere in the world, has had the same dilemma and weighed the pros and cons. You can read that conversation and those comments to help you make the same decision.

This can be applied to business and professional development as well. You can learn more about new skills, which selections or options are better for your business, and so forth. Reddit is like a personalized stream of consciousness that is happy to help you.

2. Learning Something New:

As we just mentioned, you can learn something brand new on Reddit. If you want to know more about quantum mechanics, the meaning of life, the start of the universe, or even the different critters in the ocean we've never discovered before, you can go learn about it on Reddit. There are lots of experts, news junkies, and researchers on Reddit that will present you with verifiable links that go right to the source of the information as well. Then you can read comments that expand upon the already discussed topic to really get an in-depth understanding of the new information.

3. Join a Community of People Like Yourself:

Maybe you are absolutely obsessed with Lord of the Rings, or maybe you just want to talk about the Latin words that appeared in the base of all Harry Potter spells. Your friends and family in real life make fun of you for this. But, there are other people like you in the world, and they are on Reddit, in a special subreddit where you can finally be yourself. You can interact with other people that share your passions, helping you to feel less alone in the world.

4. Fake News Proof:

Thanks to the aggressive Reddit culture of not tolerating anything that is fake, false, spammed, or self promotional, almost every piece of fake news is zapped from the site before moderators need to even do anything. With down voting, users will quickly flag any information or any links that are exaggerating the truth. Unlike other social media sites that struggle with fake news every day, Reddit is the closest thing you can use to finding the truth. There is something to be said for that.

5. AMA for Brand and Personal Interviews:

AMAs are a great way to promote products, businesses, and services without actually directly promoting them. Plus, there is something you have to offer the world of Reddit, and allowing them to ask you whatever they want will help you feel more connected to the platform. It's a great place for marketers, even if they aren't outright welcomed on there today. Test your creative marketing abilities and benefit from the traffic that is naturally occurring on Reddit today.

CHAPTER 9:

The Future of Reddit

So, where do we see Reddit going into the future? Yes, it has survived 15-years in existence, but does that mean it will always be around? These are questions every social media founder asks themselves over coffee in the morning, and rightfully so. Apps like TikTok came out of left field and are disrupting the usual social media experience we've enjoyed for some years.

Will Reddit stand the test of time?

What Makes Reddit So Unique

Reddit is its own universe, as you by now understand. Although it's responsible for so many good and bad things in the world today, here are some of the reasons why its uniqueness has fortified its continued popularity now and into the future:

- **It shapes internet culture.** Internet culture is the foundation of future generations. How we talk, share information, support each other, exchange jokes, and debate is important. Many of these norms are born right on Reddit.

- **It creates memes.** Memes have become so big that even political candidates are leveraging them today. Meme creators feel free to share their work right on Reddit.

- **It paves the way for crowdfunding.** Crowdfunding is an amazing way to help people stricken by tragedy, unforeseen circumstances, or those without proper funding to have a better life for themselves. Where better to start this movement than on the community social media platform?

- **It reshapes online interviewing.** The AMA formatting on Reddit has shaped how online interviews occur today. It's made it easier to get information out of experts and celebrities.

There are really no other competitors that come close to touching Reddit's uniqueness. With things like Instagram, there's always VSCO, Facebook, TikTok, etc. for sharing pictures and video. It's not unique to Instagram. But with Reddit, no other site is based on a boring interface that is passionately guarded by users wishing to share the truth.

Especially as fake news continues to run rampant in our lives right now, a site that eradicates any mention of fake news is something that is more relevant than ever.

Oh, and with the coronavirus pandemic, our communication is entirely digital right now. We are communicating through technology like never before – it's our lifeline to other people. Reddit sits at the intersection of a desire for human connection and universal truth. It's two things we all desire while we wait out this pandemic from home. Many of these digital trends will stick even when COVID-19 subsides, as "discussion" is something that has been part of human interaction since the beginning of mankind.

What do you think the future of Reddit will look like?

Conclusion

We hope you have enjoyed our Honor Society Reddit Roundup, provided exclusively to you here in this e-book. We believe that Reddit is a highly valuable and often under-utilized tool at everyone's disposal today, from students and graduate students, to professionals and individuals looking to know more about a certain topic. Although Instagram and Facebook dominate news headlines, that doesn't mean Reddit is any less valuable.

If you take the time to get to know the Reddit community, Redditors will welcome you in with open arms. You will find this kind of connection, learning ability, and intellectual stimulation to be highly motivating and inspiring, especially in an age of social distancing.

Best of all: it's completely free to get on Reddit and find out for yourself!

Honor Society Foundation: Reddit Consensus

We are a nonprofit, platinum rated 501(c)3 by GuideStar operation. We are on a mission to provide you with value, information, and insight related to all of your current and future endeavors. Ranked a 2019 Top Rated Nonprofit by Greatnonprofits.org, Honor Society Foundation is here to help you be the best you were meant to be.

If you have any questions about your future, getting familiar with Reddit, or using the internet to your advantage, we are here to help.

Please let us know your thoughts regarding this e-book!

References

https://www.searchenginejournal.com/social-media/reddit-guide/

https://lifehacker.com/a-beginners-guide-to-reddit-1798643829

https://mashable.com/2012/06/06/reddit-for-beginners/

https://www.dailydot.com/debug/what-is-reddit/

https://mashable.com/2014/12/03/history-of-reddit/

https://websitebuilder.org/reddit-statistics/

https://www.searchenginejournal.com/why-every-marketer-should-be-on-reddit/262239/

https://www.theodysseyonline.com/use-reddit

www.ingramcontent.com/pod-product-compliance
Lightning Source LLC
Chambersburg PA
CBHW050316220526
45465CB00005B/2021